Standard Review Plan for Transportation Packages for MOX Spent Nuclear Fuel

Manuscript Completed: September 2005
Date Published: September 2005

Prepared by
R.S. Hafner, G.C. Mok, J. Hovingh,
C. K. Syn, E.W. Russell, S.C. Keaton,
J.L. Boles, D.K. Vogt, P. Prassinos

Lawrence Livermore National Laboratory
7000 East Avenue
Livermore, CA 94550-9234

J.A. Smith, NRC Project Manager

Spent Fuel Project Office
Office of Nuclear Material Safety and Safeguards
U.S. Nuclear Regulatory Commission
Washington, D.C. 20555-0001
NRC Job Code A0291

ABSTRACT

The NRC contracted with LLNL to compile this supplement to NUREG-1617 to incorporate additional information specific to mixed uranium-plutonium oxide (MOX) fuel. This supplement provides details on package review guidance resulting from significant differences between spent nuclear fuel from irradiated LEU fuel and that from irradiated MOX fuel. The information presented is not to be construed as having the force and effect of NRC regulations (except where regulations are cited), or as indicating that applications supported by safety analyses and prepared in accordance with Regulatory Guide 7.9 will necessarily be approved, or as relieving anyone from the requirements of any pertinent regulations. The principal purpose of this supplement is to ensure the quality and uniformity of staff reviews of packagings intended for transport of MOX fuel assemblies irradiated in thermal reactors only. It is also the intent of this plan to make information about regulatory matters widely available, and improve communications between NRC, interested members of the public, and the nuclear industry, thereby increasing the understanding of the NRC staff review process. In particular, this supplemental guidance, together with NUREG-1617, assists potential applicants by indicating one or more acceptable means of demonstrating compliance with the applicable regulations.

CONTENTS

FIGURES

TABLES

ACRONYMS AND ABBREVIATIONS

boiling-water reactor	BWR
burnable poison rods	BPRs
Combustion Engineering	CE
Code of Federal Regulations	CFR
fuel grade	FG
General Electric	GE
hypothetical accident conditions	HAC
heavy metal	HM
low-enriched uranium	LEU
mixed uranium-plutonium oxide	MOX
metric tons	MT
maximum normal operating pressure	MNOP
normal conditions of transport	NCT
Nuclear Regulatory Commission	NRC
Oak Ridge National Laboratory	ORNL
power grade	PG
pressurized-water reactor	PWR
Regulatory Guide	RG
Safety Analysis Report	SAR
spent nuclear fuel	SNF
weapon grade	WG
worst-case difference	WCD

INTRODUCTION

The Standard Review Plan for Transportation Packages for Spent Nuclear Fuel (NUREG 1617)[1] provides guidance for the U.S. Nuclear Regulatory Commission (NRC) safety reviews of packages used in the transport of spent nuclear fuel (SNF) under Title 10 of the U.S. Code of Federal Regulations (CFR), Part 71 (10 CFR Part 71). It is not intended as an interpretation of NRC regulations. NUREG-1617 supplements NRC Regulatory Guide (RG) 7.9, "Standard Format and Content of Part 71 Applications for Approval of Packaging for Radioactive Material,"[2] for review of package applications. NUREG-1617 involves guidance for reviewing SNF packagings intended for transport of SNF assemblies containing low-enriched uranium oxide (LEU) fuel irradiated in thermal reactors only.

This current report is not a stand-alone document but is intended as a supplement to NUREG-1617. It is intended to provide details on package review guidance resulting from significant differences between SNF contents from irradiated LEU fuel and that from irradiated mixed uranium-plutonium oxide (MOX) fuel. Nothing contained in this document may be construed as having the force and effect of NRC regulations (except where the regulations are cited), or as indicating that applications supported by safety analyses and prepared in accordance with RG 7.9 will necessarily be approved, or as relieving any person from the requirements of 10 CFR Parts 20, 30, 40, 60, 70, or 71 or any other pertinent regulations. The principal purpose of this supplement to NUREG-1617 is to ensure the quality and uniformity of staff reviews of packagings intended for transport of MOX fuel assemblies irradiated in thermal reactors only. It is also the intent of this plan to make information about regulatory matters widely available, and improve communications between NRC, interested members of the public, and the nuclear industry, thereby increasing the understanding of the NRC staff review process. In particular, this supplemental guidance, together with NUREG-1617, assists potential applicants by indicating one or more acceptable means of demonstrating compliance with the applicable regulations.

This supplement to NUREG-1617 is organized in the same manner as NUREG-1617, and has the identical numbering of subsections as found in that document. In addition, appendices found in this supplement are labeled to allow this report to be completely merged with NUREG-1617 without needing to change any labeling. For example, NUREG-1617 has three appendices labeled A, B, and C. This supplement has two appendices labeled D and E. Appendix D summarizes the differences between thermal and radiation properties of MOX SNF and LEU SNF. Appendix E contains information on benchmark considerations for MOX SNF.

The subsection numbering within each section in NUREG-1617 is the same except for Section 8, Acceptance Tests and Maintenance Program. In the other sections, the fifth subsection is labeled Review Procedures. Review Procedures lists different review approaches for each subsection. These different review approaches in each Review Procedures subsection in this supplement are consequences of significant differences between LEU-SNF or MOX-SNF packages that potentially affect the compliance of the section of the Safety Analysis Report (SAR) in question with NRC regulations. For Section 8, the different review methods are in Subsections 8.2.4 of Acceptance Tests and 8.3.4 of Maintenance Program, both of which are also labeled Review Procedures. In addition, in Section 6 on Criticality Review, modifications are required to account for the current plans by the NRC to not allow burnup credit for pressurized water reactor (PWR) irradiated MOX SNF. These modifications are included in Subsection 6.4, on Acceptance Criteria, and within that subsection, in 6.4.8, on Burnup Credit Evaluation.

One of the potentially significant differences between LEU SNF and MOX SNF results because MOX SNF can have larger heat generation rates, larger photon emission rates, and larger neutron emission rates due to decay (see Appendix D for a discussion of these features). Another potentially significant difference between LEU SNF and MOX SNF results because plutonium in MOX (note that LEU SNF also contains plutonium that is produced from the neutron capture of ^{238}U) is a significant radiological hazard, and this can affect the leakage requirements imposed on a package. In addition, several other differences are also mentioned in some of the sections that warrant review attention.

MOX SNF comes from MOX-fresh fuel that has been irradiated in a thermal reactor. The MOX-fresh fuel can be made with plutonium having various compositions of plutonium isotopes. The U.S. DOE Standard DOE-STD-3013-2000 (herein called the 3013 Standard)[3] will be used to specify typical grades of plutonium that may be employed to make the MOX-fresh fuel. The actual plutonium compositions found in practice may not match these compositions exactly, but these grades can be considered typical for the purposes of this supplement to NUREG-1617. The 3013 Standard gives weight percents for various isotopes in various grades of plutonium. They are reproduced in the following table as representative values for typical grades of plutonium used to fabricate fresh MOX fuel.

**Table 1. Typical Isotopic Mix in Weight Percent for Various Grades of
Plutonium as Specified in the 3013 Standard**
(See Note a.)

Isotope	Weapon Grade	Fuel Grade	Power Grade
^{238}Pu	0.05	0.1	1.0
^{239}Pu	93.50	86.1	62.0[b]
^{240}Pu	6.00	12.0	22.0
^{241}Pu	0.40	1.6	12.0
^{242}Pu	0.05	0.2	3.0

[a] ^{236}Pu and ^{241}Am could be present but are not included in the 3013 Standard

[b] 63% reduced to 62% so that sum is 100%

References

1. U.S. Nuclear Regulatory Commission, "Standard Review Plan for Transportation Packages for Spent Nuclear Fuel," NUREG-1617, U.S. Government Printing Office, Washington, D.C., 2000.

2. U.S. Nuclear Regulatory Commission, "Standard Format and Content of Part 71 Applications for Approval of Packaging for Radioactive Material," Regulatory Guide 7.9, Rev. 1, 1986.

3. U.S. Department of Energy, "Stabilization, Packaging, and Storage of Plutonium-Bearing Materials," U.S. DOE Standard DOE-STD-3013-2000, Washington D.C., September 2000.

1.0 GENERAL INFORMATION REVIEW

1.5 Review Procedures

The general information review of NUREG-1617 is applicable to the review of both MOX-SNF and LEU-SNF packages. In this section, no significant deviations exist in the review procedures and considerations for the two packages. This section considers each of the subsections of Section 1.5 (Review Procedures) of NUREG-1617, and highlights the special considerations or attention needed for MOX-SNF packages. In subsections where no significant differences were found, that particular subsection has been omitted from this section.

For all packages, the general information review is based in part on the descriptions and evaluations presented in the Structural Evaluation, Thermal Evaluation, Containment Evaluation, Shielding Evaluation, Criticality Evaluation, Operating Procedures and Acceptance Tests and Maintenance Program sections of the SAR. Similarly, results of the general information review are considered in the review of the SAR sections on Structural Evaluation, Thermal Evaluation, Containment Evaluation, Shielding Evaluation, Criticality Evaluation, Operating Procedures, and Acceptance Tests and Maintenance Program.

2.0 STRUCTURAL REVIEW

2.5 Review Procedures

The structural review of NUREG-1617 is generally applicable to the review of both MOX-SNF and LEU-SNF packages. In this section, no significant deviations exist in the review procedures and considerations for the two packages. This section considers each of the subsections of Section 2.5 (Review Procedures) of NUREG-1617, and highlights the special considerations or attention needed for the MOX-SNF packages. In subsections where no significant differences were found, that particular subsection has been omitted from this section.

For all packages, the structural review is based in part on the descriptions and evaluations presented in the General Information and the Thermal Evaluation sections of the SAR. Similarly, results of the structural review are considered in the review of the SAR sections on Thermal Evaluation, Containment Evaluation, Shielding Evaluation, Criticality Evaluation, Operating Procedures, and Acceptance Tests and Maintenance Program.

3.0 THERMAL REVIEW

3.5 Review Procedures

The thermal review of NUREG-1617 is generally applicable to the review of both MOX-SNF and LEU-SNF packages. In this section, no significant deviations exist in the review procedures and considerations for the two packages. This section considers each of the subsections of Section 3.5 (Review Procedures) of NUREG-1617, and highlights the special considerations or attention needed for MOX-SNF packages. In subsections where no significant differences were found, that particular subsection has been omitted from this section.

For all packages, the thermal review is based in part on the descriptions and evaluations presented in the General Information and the Structural Evaluation sections of the SAR. Similarly, results of the thermal review are considered in the review of the SAR sections on Structural Evaluation, Containment Evaluation, Shielding Evaluation, Criticality Evaluation, Operating Procedures, and Acceptance Tests and Maintenance Program.

3.5.1 Description of the Thermal Design

3.5.1.3 Content Heat Load Specification

Although the assembly decay heat generation rate for a MOX-SNF assembly may be larger than the decay heat generation rate for LEU SNF by 20% or more for Weapons Grade (WG) plutonium in MOX, and possibly 50% to 100% greater for Fuel Grade (FG) or Power Grade (PG) plutonium in MOX (see Appendix D), there should be no significant differences in the general methods to be used for review of LEU-SNF or MOX-SNF packages.

4.0 CONTAINMENT REVIEW

4.5 Review Procedures

The containment review of NUREG-1617 is generally applicable to the review of both MOX-SNF and LEU-SNF packages. In this section, however, a few significant deviations may exist in the review procedures and considerations for the two packages. This section considers each of the subsections of Section 4.5 (Review Procedures) of NUREG-1617, and highlights the special considerations or attention needed for MOX-SNF packages. In subsections where no significant differences were found, that particular subsection has been omitted from this section.

For all packages, the containment review is based in part on the descriptions and evaluations presented in the General Information, Structural Evaluation, and the Thermal Evaluation sections of the SAR. Similarly, results of the containment review are considered in the review of the SAR sections on Operating Procedures, and the Acceptance Tests and Maintenance Program.

4.5.2 Containment under Normal Conditions of Transport

4.5.2.2 Containment Criteria

Reference is made in this subsection of NUREG-1617 to ANSI N14.5[4-1] and to NUREG/CR-6487.[4-2] With respect to the methodology and the calculations presented in NUREG/CR-6487, additional consideration will need to be given to the possibility of increased levels of plutonium isotopes inherent in MOX SNF. For calculational purposes, increased levels of plutonium in the fuel will have an influence on the mass fraction of the fuel that could be released as fines during a cladding breach. (See Section 6.1.2 and Table 6-2 of NUREG/CR-6487. See also Table 4-1 of NUREG-1617.)

With an A_2 value of 1.0×10^{-3} TBq (2.7×10^{-2} Ci) for most of the common plutonium isotopes, a relatively small increase in the plutonium-containing fines could have a significant influence on the overall containment criteria. The variation in the other parameters presented in NUREG/CR-6487 notwithstanding, a relatively small increase in the value of plutonium-containing fines for MOX SNF could easily change the normally expected leakage test criterion for LEU SNF from the 10^{-5} cm^3/sec range (currently suggested in Table 6-9 of NUREG/CR-6487) to a value that is substantially lower.

Consideration should also be given to defaulting to the "leaktight" criterion specified in ANSI N14.5 for the normally expected leakage test criterion for MOX SNF.

4.5.2.3 Compliance with Containment Criteria

Other than the comments noted above for Section 4.5.2.2, there should be no significant differences in the general methods to be used for review of LEU-SNF or MOX-SNF packages.

4.5.3 Containment under Hypothetical Accident Conditions

4.5.3.2 Containment Criteria

Comments were provided above with respect to Section 4.5.2.2 of this document. Similar comments also pertain to this section, except that the leakage test value(s) referred to from Table 6-9 of NUREG/CR-6487 should be adjusted to reflect the Hypothetical Accident Conditions values shown in the same table.

4.5.3.3 Compliance with Containment Criteria

Other than the comments noted above for Sections 4.5.2.2 and 4.5.3.2, there should be no significant differences in the general methods to be used for review of LEU-SNF or MOX-SNF packages.

References

4-1. Institute for Nuclear Materials Management, "American National Standard for Radioactive Materials — Leakage Tests on Packages for Shipment," ANSI N14.5-1997, New York, NY, 1998.

4-2. U.S. Nuclear Regulatory Commission, "Containment Analysis for Type B Packages Used to Transport Various Contents," NUREG/CR-6487, U.S. Government Printing Office, Washington, D.C., 1996.

5.0 SHIELDING REVIEW

5.5 Review Procedures

The shielding review of NUREG-1617 is generally applicable to the review of both MOX-SNF and LEU-SNF packages. In this section, however, a few significant deviations may exist in the review procedures and considerations for the two packages. This part considers each of the subsections of Section 5.5 (Review Procedures) of NUREG-1617, and highlights the special considerations or attention needed for MOX-SNF packages. In subsections where no significant differences were found, that particular subsection has been omitted from this section.

For all packages, the shielding review is based in part on the descriptions and evaluations presented in the General Information, Structural Evaluation, and Thermal Evaluation sections of the SAR. Similarly, results of the shielding review are considered in the review of the SAR sections on Operating Procedures, and on Acceptance Tests and Maintenance Program.

5.5.2 Source Specification

5.5.2.1 Gamma Source

Although the assembly decay photon emission rate for a MOX SNF assembly may be larger than the decay photon emission rate for LEU SNF by 20% or more for WG plutonium in MOX, and possibly 50% to 100% greater for FG or PG plutonium in MOX (see Appendix D), there should be no significant differences in the general methods to be used for review of LEU-SNF or MOX-SNF packages.

5.5.2.2 Neutron Source

One potential difference in the review approach between LEU SNF and MOX SNF is that the neutron dose rate can be more important relative to the gamma dose rate. In addition, the decay neutron emission rate for MOX SNF can be up to an order of magnitude larger than the decay neutron emission rate for LEU SNF (see Appendix D). This means that particular care is necessary in determining the appropriate neutron source strength. The contribution from (α,n) reactions is not necessarily small relative to spontaneous fission for MOX SNF, and can predominate. Depending on the methods used to calculate these source terms, the applicant may still determine the energy group structure independently for spontaneous fission and (α,n) reactions. However, it is generally necessary to include contributions from both spontaneous fission and (α,n) reactions.

The effect of differences in neutron energies and spectral distribution between LEU SNF and MOX SNF can become important to review approaches. The reviewer should be vigilant in identifying potentially large decay neutron emission rates for MOX SNF. Curium-244 is typically the dominant neutron emitter for times after discharge beyond about 6 months. However, ^{242}Cm can predominate for shorter times, i.e., less than about 6 months after discharge. Also, neutron multiplication effects can be important sources of additional neutrons, and should be included in the shielding analysis (most modern radiation transport codes inherently produce multiplication neutrons).

6.0 CRITICALITY REVIEW

6.4 Acceptance Criteria

6.4.8 Burnup Credit Evaluation

NRC staff does not plan to allow any credit for burnup of fissile material or increase in actinide or fission product poisons during irradiation for either MOX-BWR-SNF or MOX-PWR-SNF assemblies in the near future. The NRC staff does not currently allow any credit for burnup of LEU-BWR-SNF assemblies so the treatment for MOX-BWR-SNF assemblies will remain similar. However, burnup credit is currently allowed for LEU-PWR-SNF assemblies, so MOX-PWR-SNF assemblies will be treated differently than LEU-PWR-SNF assemblies[*]. MOX-BWR-SNF, LEU-BWR-SNF and MOX-PWR-SNF assemblies must be considered as fresh fuel for purposes of criticality safety determinations. Therefore, Subsections 6.4.8.1 through 6.4.8.6 in NUREG-1617 do not apply for MOX-SNF.

6.5 Review Procedures

The criticality review of NUREG-1617 is generally applicable to the review of both MOX-SNF and LEU-SNF packages. In this section, however, a few significant deviations may exist in the review procedures and considerations for the two packages. This section considers each of the subsections of Section 6.5 (Review Procedures) of NUREG-1617, and highlights the special considerations or attention needed for MOX-SNF packages. In subsections where no significant differences were found, that particular subsection has been omitted from this section.

For all packages, the criticality review is based in part on the descriptions and evaluations presented in the General Information, Structural Evaluation, and Thermal Evaluation sections of the SAR. Similarly, results of the criticality review are considered in the review of the SAR sections on Operating Procedures, and on the Acceptance Tests and Maintenance Program.

6.5.2 Spent Nuclear Fuel Contents

NRC staff does not currently allow any credit for burnup of the fissile material or increase in actinide or fission product poisons during irradiation for either MOX-PWR-SNF, or MOX-BWR-SNF assemblies. Criticality safety for both types of assemblies is based on fresh-fuel assemblies.

Guidelines for reviewing LEU-fresh-fuel and MOX-fresh-fuel criticality are similar since their criticality characteristics regarding moderation, absorption, and interaction are similar. Other than burnup credit treatments, there should be no significant differences in the general methods to be used for review of LEU-SNF or MOX-SNF packages.

6.5.7 Benchmark Evaluations

6.5.7.1 Experiments and Applicability

Criticality safety for either MOX-PWR-SNF or MOX-BWR-SNF assemblies is based on fresh-fuel assemblies. There are considerably fewer criticality benchmark experiments using MOX-fresh fuel than LEU-fresh fuel. Therefore, differences between the package and benchmarks may be more substantial for MOX-fresh fuel than for LEU-fresh fuel, so it may be more difficult to properly consider them. Appendix E discusses the availability of

MOX-fresh-fuel benchmarks and their important characteristics from a criticality perspective. Appendix E also discusses how a reviewer might choose a set of appropriate MOX-fresh-fuel benchmarks.

6.5.7.2 Bias Determination

Criticality safety for either MOX-PWR-SNF or MOX-BWR-SNF assemblies is based on fresh-fuel assemblies. Because of the lack of criticality benchmark experiments using MOX-fresh fuel, assigning a bias value for benchmarks may be more difficult. Appendix E discusses MOX-fresh-fuel benchmarks, and how a reviewer might determine a conservative bias value from comparisons between benchmark experiments and criticality calculations of the multiplication coefficient for those experiments. Appendix E also discusses how a reviewer might determine a conservative bias value for situations when the number of MOX-fresh-fuel benchmarks is less than desirable.

6.5.8 Burnup Credit

Burnup credit is not allowed for either LEU-BWR-SNF or MOX-BWR-SNF assemblies. There are significant differences in review approaches between how burnup is treated with LEU-PWR-SNF assemblies and MOX-PWR-SNF assemblies. Burn up credit is given for LEU-PWR-SNF fuel assemblies, but not currently for MOX-PWR-SNF fuel assemblies. MOX-PWR-SNF assemblies must be considered as fresh fuel for determining criticality safety.

7.0 OPERATING PROCEDURES REVIEW

7.5 Review Procedures

The operating procedures review of NUREG-1617 is generally applicable to the review of both MOX-SNF and LEU-SNF packages. In this section, no significant deviations exist in the review procedures and considerations for the two packages. This section considers each of the subsections of Section 7.5 (Review Procedures) of NUREG-1617, and highlights the special considerations or attention needed for MOX-SNF packages. In subsections where no significant differences were found, that particular subsection has been omitted from this section.

For all packages, the operating procedures review is based in part on the descriptions and evaluations presented in the General Information, Structural Evaluation, Thermal Evaluation, Containment Evaluation, Shielding Evaluation, and Criticality Evaluation sections of the SAR.

8.0 ACCEPTANCE TESTS AND MAINTENANCE PROGRAM REVIEW

8.2 Acceptance Tests

8.2.4 Review Procedures

The acceptance tests review of NUREG-1617 is generally applicable to the review of both MOX-SNF and LEU-SNF packages. In this section, however, a few significant deviations may exist in the review procedures and considerations for the two packages. This part considers each of the subsections of Section 8.2.4 (Review Procedures) of NUREG-1617, and highlights the special considerations or attention needed for MOX-SNF packages. In subsections where no significant differences were found, that particular subsection has been omitted from this section.

For all packages, the acceptance tests review is based in part on the descriptions and evaluations presented in the General Information, Structural Evaluation, Thermal Evaluation, Containment Evaluation, Shielding Evaluation, and Criticality Evaluation sections of the SAR.

8.2.4.4 Leakage Tests

Because the containment system of the packaging is subjected to the fabrication leakage test requirements specified in ANSI N14.5,[8-1] the acceptable leakage criterion should be consistent with that identified in the Containment Evaluation section (i.e., Chapter 4) of the SAR.

As was noted previously in Section 4.5.2.2 and 4.5.3.2, there could be a significant difference in the acceptance leakage test criterion from that which normally may be expected for LEU SNF. The potential for the difference stems from the increased levels of plutonium isotopes inherent in MOX SNF. Depending on how the calculations are performed, the inherent increase in the levels of the plutonium isotopes could easily change the normally expected leakage test criterion for LEU SNF from the 10^{-5} cm^3/sec range to a value that is substantially lower for MOX SNF. (See Section 4.5.2.2, above, and Section 8.3.4.2, below.)

8.3 Maintenance Program

8.3.4 Review Procedures

The maintenance program review of NUREG-1617 is generally applicable to the review of both MOX-SNF and LEU-SNF packages. In this subsection, however, a few significant deviations may exist in the review procedures and considerations for the two packages. This part considers each of the subsections of Section 8.3.4 (Review Procedures) of NUREG-1617, and highlights the special considerations or attention needed for MOX-SNF packages. In subsections where no significant differences were found, that particular subsection has been omitted from this section.

For all packages, the maintenance program review is based in part on the descriptions and evaluations presented in the General Information, Structural Evaluation, Thermal Evaluation, Containment Evaluation, Shielding Evaluation, and Criticality Evaluation sections of the SAR.

8.3.4.2 Leakage Tests

Because the containment system of the packaging is subjected to the maintenance leakage requirements and/or the periodic leakage test requirements specified in ANSI N14.5, the acceptable leakage criterion should be consistent with that identified in the Containment Evaluation section (i.e., Chapter 4) of the SAR.

As was noted previously in Sections 4.5.2.2, 4.5.3.2, and 8.2.4.4, there could be a significant difference in the acceptance leakage test criterion from that which normally may be expected for LEU SNF. The potential for the difference stems from the increased levels of plutonium isotopes inherent in MOX SNF. Depending on how the calculations are performed, the inherent increase in the levels of the plutonium isotopes could easily change the normally expected leakage test criterion for LEU SNF from the 10^{-5} cm^3/sec range to a value that is substantially lower for MOX SNF. (See Section 4.5.2.2 and Section 8.2.4.4, above.)

References

8-1. Institute for Nuclear Materials Management, "American National Standard for Radioactive Materials — Leakage Tests on Packages for Shipment," ANSI N14.5-1997, New York, NY, 1998.

APPENDICES

APPENDIX D:
DIFFERENCES BETWEEN THERMAL AND RADIATION PROPERTIES OF MOX AND LEU SPENT NUCLEAR FUEL

This appendix reviews the expected differences between thermal and radiation properties of MOX and LEU spent nuclear fuel (SNF). Limited experimental information is available for MOX SNF, so determining what to expect from various grades of plutonium (see below), assembly types, fuel pellet types, reactor categories, and amount of burnup is determined solely from performing source term calculations. While only limited studies have been performed to understand what might be expected from these types of variations, educated estimates for these differences are attempted here, and noted in the text or in footnotes.

Oak Ridge National Laboratory (ORNL) conducted a detailed study of the rates of heat generation, gamma emission, and neutron emission due to decay for MOX fuel irradiated in various reactors. Four ORNL reports present the results for SNF from the following reactors:

1. Combustion Engineering (CE) System 80+ pressurized-water reactor (PWR) design.[D-1] This report gives the results for both MOX fuel and LEU fuel. The assessment given for MOX-fuel and LEU-fuel assemblies will be used as our generic fuel comparisons for PWRs.

2. General Electric (GE) boiling-water reactor (BWR) design.[D-2] This report gives the results for both MOX fuel and LEU fuel. The assessment given for MOX- and LEU-fuel assemblies will be used as our generic fuel comparisons for BWRs.

3. Westinghouse PWR design.[D-3]

4. Westinghouse PWR design without integral absorbers.[D-4]

For each reactor type, it is possible to (1) select from a number of different fuel assemblies, (2) for MOX, choose different arrangements of fuel pins having different compositions of plutonium, uranium, and burnable absorbers, and (3) use annular fuel pellets rather than cylindrical fuel pellets. All of these changes can affect the total burnup, and the amount of heavy metal contained in the MOX-fuel or LEU-fuel assemblies. The ORNL studies focused on identifying differences in spent fuel characteristics that are significantly greater than typical burnup-related variations. It is expected that increasing the burnup of both MOX-fuel and LEU-fuel assemblies would result in larger differences in spent fuel characteristics. The two ORNL reports were chosen for this study because they consider typical differences in SNF characteristics, and they are the only ones available that compare LEU SNF to MOX SNF. They do not necessarily, however, represent analyses that give bounding differences in SNF characteristics.

The ORNL reports used Weapons Grade (WG) plutonium for their MOX-fuel rods. The 3013 Standard[D-5] gives weight percents for various isotopes in various grades of plutonium (see Table 1). The ORNL reports used weight percents of various plutonium isotopes consistent with those for WG listed in Table 1 in the text. Details of the fuel assemblies used in the ORNL studies are presented below.

A discussion of the characteristics of the CE System 80+ PWR-MOX-SNF and PWR-LEU-SNF fuel assemblies is presented below. For purposes of comparison, these same data are also summarized below, in Table D-1. The irradiation characteristics of the CE System 80+ fuel assemblies are also shown below, in Table D-2. A comparison of fuel assembly characteristic of the GE BWR-MOX-SNF and BWR-LEU-SNF fuel assemblies is shown below, in Table D-3. The irradiation characteristics of the GE BWR fuel assemblies are shown below, in Table D-4.

The MOX fuel for the CE-PWR irradiation contained 6.7 wt% WG plutonium and 91.3 wt% depleted uranium,* together with 1.9 wt% of erbium, in the form of Er_2O_3, as components of the heavy metal (HM). The core also contained Al_2O_3-B_4C burnable poison rods (BPRs). The assembly studied, known as the shim assembly, contained a 16×16 square array that was 20.25 cm on a side, with a fuel-rod pitch of 1.29 cm. The assembly studied contained 256 fuel rod positions with a total of 224 fuel rods, four (4) control rods, one (1) instrument tube, and 12 BPRs. The four control rods and single instrument tube displaced the equivalent of 20 fuel rod positions. The assembly contained 0.419 metric tons (MT) of HM in the 224 fuel rods, not counting the 1.9 wt% of erbium. The burnup criterion used was 28.9 MW/MTHM, and the assembly was burned to 17,681.8 MWd, in four cycles of 365 days each. A 30-day downtime was allowed between cycles. This represents an assembly power level of 12.34 MW, and a burnup of 42.2 GWd/MTHM. (See Table D-1.)

The LEU fuel for the CE-PWR irradiation contained 4.2 wt% ^{235}U and 95.8 wt% ^{238}U, as components of the HM, in 224 identical fuel rods. In addition, 12 fuel rods contained 4.1 wt% ^{235}U and 94.0 wt% ^{238}U, together with 1.9 wt% of erbium, in the form of Er_2O_3, as components of the HM. These 12 fuel rods were located in the same positions where the 12 BPRs were located in the MOX case, above. The shim assembly studied contained the same 16×16 square array that was 20.25 cm on a side, with a fuel-rod pitch of 1.29 cm. The assembly studied also contained the same four (4) equivalent control rods, and 1 equivalent instrument tube, as the MOX assembly. The assembly contained 0.424 MT of HM in the 236 fuel rods, not counting the same 1.9 wt% of erbium. The burnup criterion used was 29.1 MW/MTHM, and the assembly was burned to 20,267.2 MWd, in three cycles of 18 months each. A comparable 30-day downtime was allowed between cycles. This represents an assembly power level of 12.34 MW, which was the same as for the MOX-fuel assembly. The burnup was 47.8 GWd/MTHM. (See Table D-1.)

Table D-1. Comparison of Fuel Assembly Characteristics for the Combustion Engineering System 80+ Pressurized-Water-Reactor SNF

Characteristic	CE-PWR MOX	CE-PWR LEU
Weight Heavy Metal (MT)	0.419	0.424
wt% WG Plutonium	6.7	NA
wt% uranium*	91.3	100 (4.2 ^{235}U)
wt% erbium (Er_2O_3)	1.9	1.9
Burnable Poison Rods (BPRs) material	Al_2O_3-B_4C	NA
Array size	16×16 (20.25 on side)	16×16 (20.25 on side)
Fuel Rod Pitch (cm)	1.29	1.29
Number of Rods	256	256
Fuel rods	224	236
Control rods	4	4 (equivalent)
Instrument tubes	1	1
BPRs rods	12	NA
Burnup Criterion (MW/MTHM)	28.9	29.1
Burnup (MWd)	17681.8	20267.2
Cycles / length	4/365 days each	3/18 months each
Assembly Power Level (MW)	12.34	12.34
Representative Burnup (GWd/MTHM)	42.2	47.8

Table D-2. **Irradiation Characteristics of Combustion Engineering System 80+ Pressurized-Water-Reactor SNF**

Fuel Type	MTHM	Irradiation (days)	Burnup (GWd/MTHM)
MOX	0.419	1,460	42.2
LEU	0.424	1,620	47.8

The MOX fuel for the GE-BWR-5[†] irradiation contained 2.97 wt% WG plutonium, 96.50 wt% depleted uranium, and 0.53 wt% gadolinium, as components of the HM. The assembly studied contained an 8×8 array that was 15.24 cm on a side, with a fuel-rod pitch of 4.129 cm. The assembly contained 64 fuel rod positions, with a total of 60 fuel rods, and one guide tube. Seven different types of fuel rods were used, each having a different amount of plutonium, uranium, and gadolinium. The guide tube displaced the equivalent of 4 fuel rod positions. The assembly contained 0.179 MT of HM, in the 60 fuel rods, not counting the 0.53 wt% of gadolinium. The burnup criterion used was 25.5 MW/MTHM, and the assembly was burned to 6,715.4 MWd, in four cycles of 340-day uptime, and a 113-day downtime, each with an additional final 113-day uptime. This amounted to an assembly power level of 4.610 MW, and a burnup of 37.6 GWd/MTHM. (See Table D-3.)

Table D-3. **Comparison of Fuel Assembly Characteristic for the General Electric Boiling-Water-Reactor SNF**

Characteristics	GE-BWR MOX	GE-BWR LEU
Weight Heavy Metal (MT)	0.179	0.183
wt% WG Plutonium	2.97	NA
wt% uranium[*]	96.50	100 (3.25 ^{235}U)
wt% gadolinium (Gd_2O_3)	0.53	2.17
Array size	8×8 (15.24 cm on side)	8×8 (15.24 cm on side)
Fuel Rod Pitch (cm)	4.129	4.129
Number of Rods	64	64
Fuel rods	60	60
Guide tube	1	1
Burnup Criterion (MW/MTHM)	25.5	25.5
Burnup (MWd)	6,715.4	6,880.8
Cycles / length	4/340-days uptime, 113-day downtime, with an additional 113-day uptime, each	4/340-days uptime, 113-day downtime, with an additional 113-day uptime, each
Assembly Power Level (MW)	4.610	4.724
Representative Burnup (GWd/MTHM)	37.6	37.6

The LEU fuel for the GE-BWR-5[†] irradiation contained 3.25 wt% ^{235}U and 96.75 wt% ^{238}U, as components of the HM, in 56 identical fuel rods. In addition, four (4) fuel rods with burnable absorbers were used, containing 2.17 wt% of Gd_2O_3. The assembly studied contained the same 8×8 array that was 15.24 cm on a side, with a fuel-rod pitch of 4.129 cm. The assembly studied contained a total of 60 fuel rods and one guide tube. The assembly contained 0.183 MT of HM in the 60 fuel rods, not counting the 2.17 wt% of gadolinium. The burnup criterion used was 25.5 MW/MTHM, which was the same as for the MOX fuel assembly. The assembly was burned to

6,880.8 MWd, in four cycles of 340-day uptime, and a 113-day downtime, each with an additional final 113-day uptime. This amounted to an assembly power level of 4.724 MW, and an identical burnup of 37.6 GWd/MTHM. (See Table D-3.)

Table D-4.   Irradiation Characteristics of General Electric Boiling-Water-Reactor SNF

Fuel Type	MTHM	Irradiation (days)	Burnup (GWd/MTHM)
MOX	0.1786	1,473	37.6
LEU	0.183	1,473	37.6

The ratios for heat generation rates, photon emission rates, and neutron emission rates vs. time-from-discharge for the CE PWR fuel assemblies are shown below in Figures D-1, D-2, and D-3, respectively. The data presented in these figures were calculated by taking the calculated rates of heat generation, gamma emission, and neutron emission due to decay for the MOX-fuel assembly irradiation and dividing them by the similar quantities for the LEU-fuel assembly. The differences in calculated decay rates for these quantities for the MOX-fuel assembly irradiation and the LEU-fuel assembly irradiation in a PWR are attributed primarily to differences in fuel material for the purposes of this study.

The ratios for heat generation rates, photon emission rates, and neutron emission rates vs. time-from-discharge for the GE BWR fuel assemblies are also shown in Figures D-1, D-2, and D-3, respectively. Again, the data presented in these figures were calculated by taking the calculated rates of heat generation, gamma emission, and neutron emission due to decay for the MOX-fuel assembly irradiation and dividing these by the similar quantities for the LEU-fuel assembly. And, again, the differences in calculated decay rates for these quantities for the MOX-fuel assembly irradiation and the LEU-fuel assembly irradiation in a BWR are attributed primarily to differences in fuel material for the purposes of this study.

Figure D-1 for heat generation rate shows that the heat rate generated by the MOX SNF and LEU SNF is within about 15% of each other over a period of 10 years after discharge. Figure D-2 for decay gamma emission rate, where only gamma energies greater than 250 keV are included in the curves,[*] shows that the decay gamma emission rate generated by the MOX SNF and LEU SNF are also within about 15% of each other over a period of 10 years after discharge. Figure D-3 for decay neutron emission rate shows that the decay neutron emission rate generated by the MOX SNF and LEU SNF differs by up to about a factor of 2.5 over a period of 10 years after discharge.[+] These results are based on a single assembly type and fuel composition for each of the two categories of reactors studied. Weapons Grade plutonium was used for both studies.

---

[*]   The shielding associated with SNF packagings is expected to absorb essentially all gammas with energies less than 250 keV.

[+]   The two curves in Figure D-3 are based on very different fuels and burnups, and the differences are not due primarily to the reactor type (PWR vs. BWR) as might be interpreted from the figure. In the BWR case, the LEU fuel has a fissile content of 3.25 wt% ^{235}U, while the MOX fuel has a fissile content of 3.1 wt%, or about 5% less. The burnups in both instances were identical. This rough equality means a MOX-to-LEU neutron emission ratio of about two is representative of the difference between LEU and MOX fuels under roughly the same reactor conditions. In the PWR case, the LEU fuel has a fissile content of 4.2 wt% ^{235}U, while the MOX fuel has a total fissile content of about 6.9 wt% (^{239}Pu+^{240}Pu+^{235}U), or about 64% more. The slightly lower (13%) burnup of the MOX fuel, combined with the higher fissile content (64%), results in the LEU fuel being significantly more burned (relative to its fissile content) compared to the MOX fuel. Burnup is based on the total heavy metal content, not fissile content, and so differences in fissile content effectively translate to differences in burnup. The neutron source, in particular, is very sensitive to burnup (or fissile content). This burnup dependence probably affects the PWR curves in Figures D-1 and D-2 also, although not to the extent of the neutron source.[D-8]

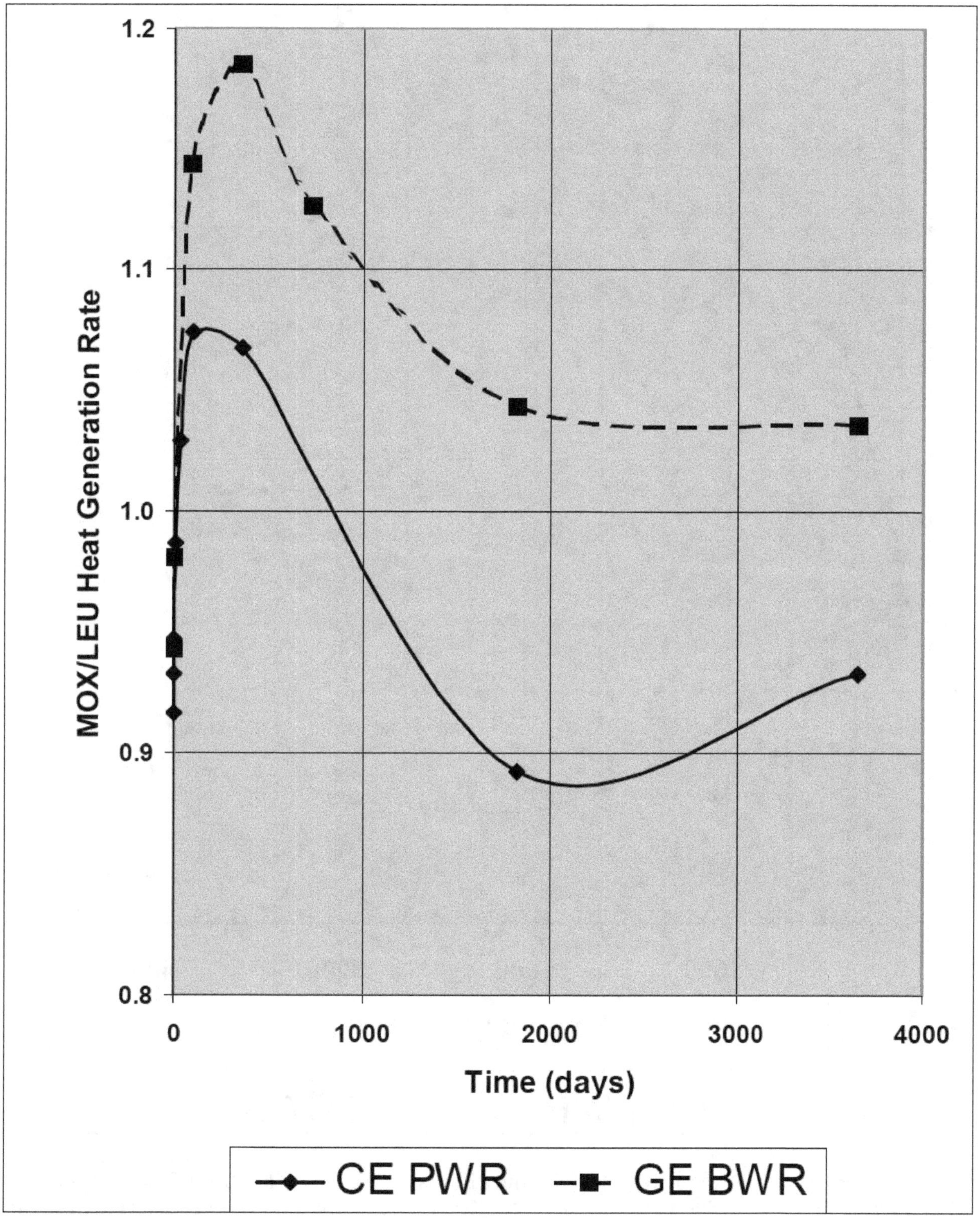

Figure D-1. Ratio of MOX to LEU Decay Heat Generation Rate vs. Time-from-Discharge for Combustion Engineering System 80+ Pressurized Water Reactor (CE-PWR) and General Electric Boiling Water Reactor Model 5 (GE-BWR).

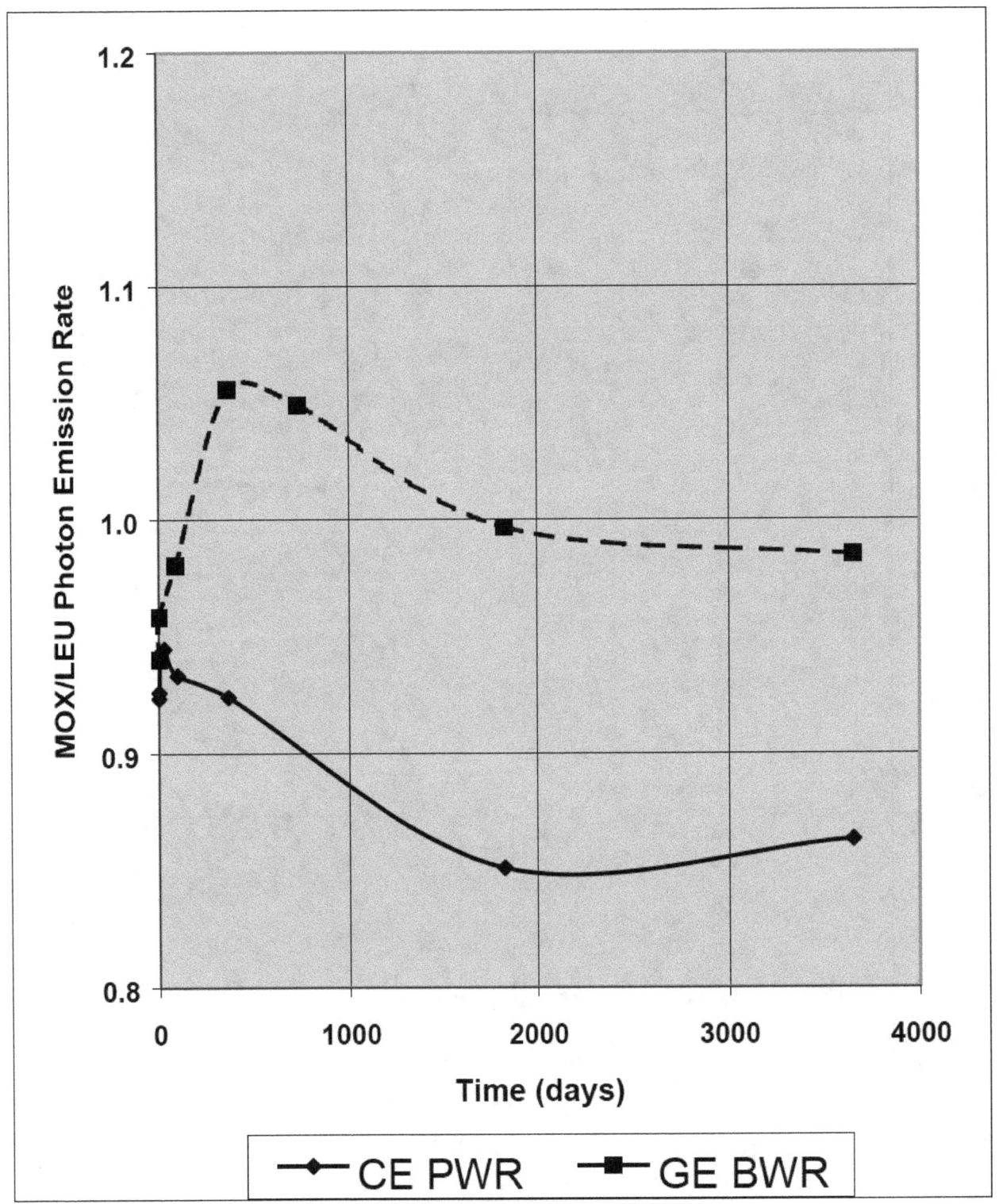

Figure D-2.   Ratio of MOX to LEU Decay Gamma Emission Rate vs. Time-from-Discharge for
Combustion Engineering System 80+ Pressurized Water Reactor (CE-PWR) and General Electric Boiling
Water Reactor Model 5 (GE-BWR).
(Note: Only gamma energies greater than 250 keV are included in these curves.)

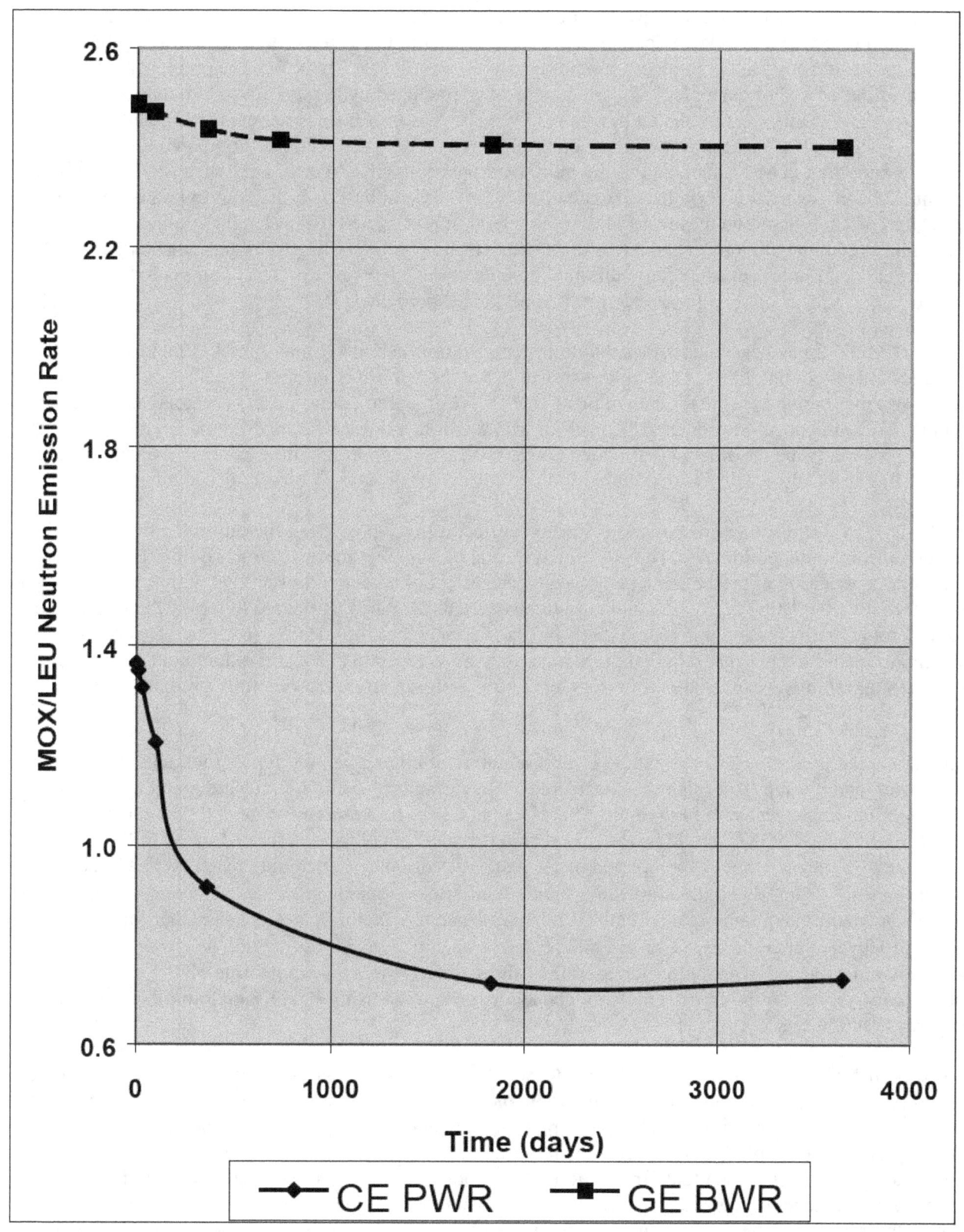

Figure D-3. Ratio of MOX to LEU Decay Neutron Emission Rate vs. Time-from-Discharge for Combustion Engineering System 80+ Pressurized Water Reactor (CE-PWR) and General Electric Boiling Water Reactor Model 5 (GE-BWR).

Most of the benchmarking that ORNL has investigated for decay heat and radiation source terms has involved LEU fuel. Limited MOX benchmarks indicate that the predicted actinide concentrations, particularly the fissile plutonium isotopes and many fission products, are not nearly as accurate for MOX fuels as previously observed for commercial LEU fuels. For example, ^{239}Pu tends to be over-predicted by about 10% to 50% and americium isotopes are also significantly over-predicted by about 25%. The reasons for this are not entirely clear, but could be due to larger uncertainties in the plutonium and other higher actinide cross sections (compared to uranium) which are more important in MOX fuel, and/or the more heterogeneous MOX cores, i.e., when MOX assemblies with different HM compositions are irradiated together with LEU assemblies. It is difficult to know the accuracy for decay heat predictions based on these results, but in general, it is expected that at longer cooling times where actinides dominate, code predictions may overestimate decay heat by potentially 10-20% or more for MOX SNF based on the calculated plutonium and americium nuclide inventories. However, several dominant decay heat nuclides important at shorter cooling times are significantly under-predicted.[D-6]

The accuracy of MOX decay heat calculations would apparently be much lower than for LEU fuels, but may be conservative for longer cooling times, and non-conservative for short cooling times. For neutron source terms, comparisons with the limited benchmark data indicate that SCALE[D-7] predictions are in very good agreement for MOX fuel in a PWR, but are over-predicted (~20%) for MOX fuel in a BWR.[D-8] Uncertainties in the computational predictions by the amounts estimated above (10-20%) support the differences in the values shown in Figures D-1 through D-3.

The use of ENDF/B-IV and earlier cross sections have the effect of over-predicting the multiplication coefficient, k_{eff}, for materials containing plutonium. The use of newer, and presumably more accurate, ENDF/B-V and VI cross sections do a better job of predicting k_{eff}. However, the effect of newer cross sections will not necessarily be less conservative for calculating decay heat and radiation source terms when compared to the earlier ones. The much larger isotopic biases observed for MOX fuels in limited benchmark studies are likely to translate into higher uncertainties (biases) in aggregate fuel properties; they may translate to a lesser extent than the isotopic analyses might suggest due to cancellation of errors, e.g., bulk fuel properties are generally predicted better than individual isotopic analyses.[D-8]

Are the studies shown in Figures D-1 through D-3 representative of other assembly types, fuel pellet types, reactor categories, and burnups that might be considered? The decay heat emission rate and gamma emission rate using WG plutonium are expected to be similar. That is, quantities of heat emission rate and gamma emission rate for MOX SNF and LEU SNF should be roughly within the same envelope determined in the ORNL studies, i.e., within about 40% of each other over a period of 10 years after discharge, including benchmark and cross section uncertainties.* Using WG plutonium, we estimate (since no systematic studies have been performed as yet) that the decay neutron emission rate for MOX SNF may be up to a factor of 4 larger than that for LEU SNF over a period of 10 years after discharge, taking into account benchmark and cross section uncertainties.+ The uncertainties are not expected to apply to shorter cooling times, relative to a discharge time of 10 years. The differences, as always, need to be confirmed by independent verification using established radiation transport codes, and cross-section sets.

Will these relationships change when studies are made with MOX fuel produced with Fuel Grade (FG) or Power Grade (PG) plutonium? The answers for heat generation and gamma emission rates due to decay are expected to be similar, but differ by a larger amount. The use of plutonium containing less ^{239}Pu and more of other plutonium isotopes means larger masses of plutonium might be required in the fuel rods, which increases the amount of other isotopes of plutonium in MOX-fresh fuel. Irradiation of fuel rods containing more of the other plutonium isotopes is expected to generate a greater heat generation rate and to emit a greater decay gamma emission rate than the WG plutonium used in the MOX fuel studied in the ORNL reports. The additional amount of other plutonium isotopes is expected to generate greater heat generation and gamma emission rates due to decay after

---

*   This is just an opinion, since the uncertainty may be larger than the increase estimated by 20%×2 = 40%.

irradiation. The heat generation and gamma emission rates due to decay for MOX SNF and LEU SNF might be within about 100% of one another over a period of 10 years after discharge, including benchmark and cross section uncertainties, although without systematic studies this is just an estimate. The uncertainties are not expected to apply to short cooling times relative to a time after discharge of 10 years.

For decay neutron emission rates, it may be more difficult to determine the amount of increase that might be expected with MOX fuel produced with another grade of plutonium, since no systematic studies have been performed as yet. The decay neutron emission rates from other grades of plutonium can be two to four times larger than those for WG plutonium. Again, the use of plutonium containing less ^{239}Pu and more of other plutonium isotopes means larger masses of plutonium might be required in the fuel rods, which increases the amount of the other isotopes of plutonium in MOX fresh fuel. Irradiation of fuel rods containing more of other plutonium isotopes is expected to generate a greater decay neutron emission rate than the WG plutonium used in the MOX fuel studied in the ORNL reports. MOX fuel produced with PG plutonium has considerably more ^{241}Pu present in the fresh fuel. Americium-241 is produced by beta decay of ^{241}Pu with a half life of 14.4 years. For times after discharge less than a year, neutrons from ^{242}Cm and ^{244}Cm can predominate after discharge for several months or so, after which the neutrons from ^{242}Cm decrease significantly. Neutrons from ^{240}Pu, and ^{241}Am may also become significant. The neutron emission rates for MOX SNF and LEU SNF should be within an order of magnitude of one another over a period of 10 years after discharge, including benchmark and cross section uncertainties, although without systematic studies this is just an estimate. The uncertainties are not expected to apply to short cooling times relative to a time after discharge of 10 years. The differences, as always, need to be confirmed by independent verification using established radiation transport codes.

References

D-1. B. D. Murphy, "Characteristics of Spent Fuel from Plutonium Disposition Reactors, Vol. 1: The Combustion Engineering System 80+ Pressurized-Water Reactor Design," Oak Ridge National Laboratory, ORNL/TM-13170/V1, June 1996.

D-2. J. C. Ryman and O.C. Hermann, "Characteristics of Spent Fuel from Plutonium Disposition Reactors, Vol. 2: A General Electric Boiling-Water Reactor Design," Oak Ridge National Laboratory, ORNL/TM-13170/V2, April 1998.

D-3. B. D. Murphy, "Characteristics of Spent Fuel from Plutonium Disposition Reactors, Vol. 3: A Westinghouse Pressurized-Water Reactor Design," Oak Ridge National Laboratory, ORNL/TM-13170/V3, July 1997.

D-4. B. D. Murphy, "Characteristics of Spent Fuel from Plutonium Disposition Reactors, Vol. 4: Westinghouse Pressurized-Water Reactor Fuel Cycle Without Integral Absorber," Oak Ridge National Laboratory, ORNL/TM-13170/V4, April 1998.

D-5. U.S. Department of Energy, "Stabilization, Packaging, and Storage of Plutonium-Bearing Materials," US DOE Standard DOE-STD-3013-2000, Washington D.C., September 2000.

D-6. B. D. Murphy and R. T. Primm III, "Prediction of Spent MOX and LEU Fuel Composition and Comparison with Measurements," *Proceedings of the 2000 ANS International Topical Meeting on Advances in Reactor Physics and Mathematics and Computation into the Next Millennium*, PHYSOR2000, Pittsburgh, PA, May 7-11, 2000.

D-7. SCALE: A Modular Code System for Performing Standardized Analyses for Licensing Evaluations, NUREG/CR-0200, Rev. 4 (ORNL/NUREG/CSD-2/R4), Vols. I, II, III, October 1995. Various versions of SCALE are available.

D-8. Personal communication to Stewart C. Keeton, Physicist in Radioactive Material Transportation and Storage Projects, Fission Energy and Systems Safety Program, Lawrence Livermore National Laboratory, Livermore, CA, from Ian C. Gauld, Nuclear Engineering Applications Engineer in Computational Physics and Engineering Division, Oak Ridge National Laboratory, Oak Ridge, Tennessee, August and December 2001, and January 2002.

APPENDIX E:
BENCHMARK CONSIDERATIONS FOR MOX SPENT NUCLEAR FUEL

E.1 Experimental Benchmarks

NRC staff does not plan to allow any credit for burnup of fissile material or increase in actinide or fission product poisons during irradiation for either MOX-BWR-SNF or MOX-PWR-SNF assemblies in the near future. Therefore, benchmark criticality analyses must be performed using fresh-fuel assemblies.

Substantial guidance on how to select an appropriate set of criticality benchmark experiments for LEU fissile systems is given in NUREG/CR-5661 and in NUREG/CR-6361.[E-1, E-2] Considerably fewer benchmark experiments exist for MOX than for LEU, however. As a consequence, the guidance provided in NUREG/CR-5661 and/or in NUREG/CR-6361 cannot be applied directly to the evaluation of MOX fissile systems. The benchmarks needed for the criticality analyses of MOX packages are in the thermal energy range. This condition results because, for essentially all types of MOX, the most reactive configuration is a flooded containment.

As an alternative, the International Handbook of Evaluated Criticality Safety Benchmark Experiments[E-3] (IHECSBE) has 11 evaluated thermal-energy studies involving MOX fuel pins in various lattice experiments, and 5 evaluated thermal-energy studies involving MOX liquids in tank experiments. These can be divided into 18 sets of experiments involving different fissile oxide compositions and configurations in lattices, and 13 sets of experiments involving different liquid fissile nitrate compositions and configurations in tanks. The total number of essentially different experiments is 131. Other benchmark experiments are available throughout the world, but are not as readily available, and the vast majority have not been rigorously evaluated in the manner of those found in the IHECSBE, and are consequently of limited use for benchmark criticality analyses for MOX packages. More evaluated MOX thermal benchmarks are expected in future editions of the IHECSBE.

The 18 sets of experiments involving fissile oxides in lattices and 13 sets of experiments involving fissile nitrate liquids in tanks have been organized and shown in Tables E-1 through E-5. The various tables are separated on two features. The first is between lattice and tank experiments, and the second is on weight percent of plutonium to total plutonium plus uranium, Pu/(Pu+U). Table E-1 has lattice experiments with Pu/(Pu+U) to 5%. Table E-2 has lattice experiments with Pu/(Pu+U) from 5% to 15%. Table E-3 has lattice experiments with Pu/(Pu+U) greater than 15%. Table E-4 has tank experiments with Pu/(Pu+U) to 31% (there are no experiments with Pu/(Pu+U) less than 22%). Table E-5 has tank experiments with Pu/(Pu+U) greater than 31%. Lists of meaningful, experimental characteristics are recorded for each set of experiments together with characteristics of their corresponding computational evaluations.

Experimental plutonium benchmarks should also be taken into account as part of the initial set of benchmark experiments to be considered for a MOX package application. About four times as many thermal-plutonium-tank-liquid benchmarks exist in the IHECSBE as thermal-MOX-tank-liquid benchmarks. However, fewer thermal-plutonium-lattice benchmarks exist in the IHECSBE as thermal-MOX-lattice benchmarks.

E.2 Summary of Bias and Uncertainty Evaluation

There are two measures of the accuracy of an experiment and its associated calculation. The first measure is the effective bias (Eff-Bias) between calculation and benchmark experiment. The multiplication coefficient for a fissile system is designated as k_{eff}. Designate the calculated k_{eff} for the benchmark experiment as k_{calc} and the benchmark experimental k_{eff} as k_{exp}. If the calculational bias, β, is defined as $\beta = k_{calc} - k_{exp}$, then a quantity Δk^* can be defined as

$$\Delta k = \left(\begin{array}{ll} \beta & \text{if } k_{calc} \leq k_{exp} \\ \\ 0 & \text{if } k_{calc} > k_{exp} \end{array} \right).$$ (E-1)

For a given experimental benchmark set, Δk_{max} is chosen as the largest absolute value of the Δk given by Equation E-1 for all experiments in the set. The 95% confidence limit of k_{calc} is k_{calc} plus twice the calculated standard deviation, which is designated by 2σ. The Eff-Bias value is then given by

$$\text{Eff-Bias} = \Delta k_{max} - 2\sigma.$$ (E-2)

Eff-Bias, as defined here, is always *less* than zero. If k_{calc} is greater than k_{exp} for all experiments in a set, the Eff-Bias value is just the negative of twice the calculated standard deviation.

The second measure is the total experimental uncertainty (Exp-Uncer) that was determined by the evaluator after assessing all sources of uncertainty for the experiments in a set.[*] A worst-case difference between k_{calc} and k_{exp} can be assigned as the difference of the total experimental uncertainty and the effective bias (Exp-Uncer - Eff-Bias) for the experimental set in question. This worst-case difference (WCD), as defined here, is always *greater* than zero. It represents the upper limit of the inherent uncertainties in the ability of the computer code, together with the cross-section set used, to accurately determine the k_{eff} of a critical benchmark experiment. Therefore, a bounding multiplication coefficient, k_{safe}, at the 95% confidence limit, can be chosen to be equal to 0.95 minus WCD, where an administrative margin of safety of 0.05 has been included.[+]

Values for the variable WCD for each experimental set vary between 0.0071 to 0.0192 (0.71% to 1.92%), 0.0043 to 0.0328 (0.43% to 3.28%), 0.0023 to 0.0138 (0.23% to 1.38%), 0.0044 to 0.0180 (0.44% to 1.80%), and 0.0044 to 0.0150 (0.44% to 1.50%), for the experimental sets in Tables E-1, E-2, E-3, E-4, and E-5, respectively. No particular correlation seems to exist between WCD and the lattice configuration or pitch. Neither does there seem to be a correlation with plutonium composition type (Pu type). The plutonium composition types are given in Table 1 in the text, and are designated as weapons grade (WG), fuel grade (FG), and power grade (PG).

The maximum value for WCD found in the five tables is 0.0328 or 3.28% in k_{eff}. How accurately a criticality computer code can predict the critical value for a criticality experiment depends on the methodology employed by the code and the cross-section set used, together with the detail to which the experimental system is modeled in the computer. In addition, the basic experimental uncertainty limits the ultimate prediction accuracy possible. Of particular importance is the cross-section set. Values for WCD in the five tables that are significantly less than 0.0100 are due to the fact that k_{calc} is greater than k_{exp}. Therefore, the value for Eff-Bias, in that case, is just the negative of twice the calculated standard deviation, which is approximately 0.0020. Cross-section sets prior to ENDF/B-V over-predict plutonium reactivity, and this represents some of the reason for the over-prediction for k_{calc} for these experiments. Values for k_{safe} are not expected to be much above 0.93, except when it can be demonstrated that the criticality code and cross section set overestimates the reactivity of the MOX contents.

---

[*]   As defined in Equation E-1, Δk is always less than or equal to zero, and is consistent with the bias, $\bar{\beta}$, defined in Reference E-1. Typically, a calculational method is termed to have a negative bias if it under-predicts the critical condition.

[*]   The evaluator included sources of experimental bias or error in each k_{exp}. This does not represent an uncertainty, and so is not included in the value for total experimental uncertainty.

[+]   If the benchmarks are applied to a package application where there is a lack of experimental data, the 0.05 administrative margin may not be sufficient, and the reviewer needs to be aware of this issue. In reality, the 0.05 margin should be sufficient, but there needs to be an assessment of the adequacy of the 0.05 to establish the basis. Guidance for deciding

Analyzing an acceptable number of MOX benchmarks is the preferred way to obtain a bias value for the MOX contents of a package. With the relatively limited number of MOX critical experiments available for use in validation exercises, it is important to determine that the application of interest to the reviewer fits within the area of applicability for the set of critical benchmark experiments selected for validation. Guidance on how to select an appropriate set of benchmark experiments for a fissile system is given in NUREG/CR-5661 and in NUREG/CR-6361. An important advancement using computational methodology to select an appropriate set of benchmark experiments for a fissile package application is currently being developed for SCALE.[E-4, E-5, E-6]

A set of sensitivity and uncertainty analysis tools are being developed for version 5 of SCALE that gives a measure of the similarity of the reactivity of a package application to that of an experimental benchmark. Sensitivity coefficients for both systems are computed and give the sensitivity of each system's k_{eff} to the cross section data. These sensitivity coefficients are determined for each energy group in the cross section library chosen in the analysis, as well as the sum over all energy groups. Two integral parameters for the combined systems are produced from the sensitivity data to determine system-to-system similarities. The first parameter can be used as a gauge of system similarity to sensitivity-only. The second parameter can be used as a measure of the similarity of the systems in terms of uncertainty, not just sensitivity. The pair of integral parameter values is determined for every potential benchmark experiment with the package application of interest. When two systems produce a value of 0.8 for either integral parameter, or both, this indicates the k_{eff} response is similar enough that one system serves well to validate the criticality safety parameters for the other system. The benchmark experiments chosen for complete validation are those with high integral parameter values.[E-4, E-5, E-6]

New parameters can also be constructed from the components of the integral parameters and can be used to explore the sensitivity of specific nuclide reactions of benchmark experiments with the package application of interest. For example, if low integral parameter values are found for an application with all benchmark experiments chosen for validation, the new parameters could serve to identify which nuclides would require additional experimental benchmark data for complete validation. Also, in the validation of shipping casks for commercial fuel, numerous benchmark experiments might serve to validate the fission reactions, and thus high integral parameter values would be found. However, the new parameters could be used to find benchmarks to ensure that any poison materials in the cask are also well validated by the benchmarks. Once these sensitivity and uncertainty analysis tools are released with version 5 of SCALE, the criticality safety analyst will have a powerful set of tools to perform detailed quantitative analyses to determine the applicability of benchmark experiments to help design package applications under consideration.[E-4, E-5, E-6]

Table E-1. Important Characteristics of Lattice Experiments with Weight Percent of Pu/(Pu+U) to 5% (from IHECSBE)

	MCT-009	MCT-002	MCT-002	MCT-006	MCT-007	MCT-008	MCT-004	MCT-005
Designation for experiments[a]								
Facility where experiments conducted	Hanford	Hanford	Hanford	Hanford	Hanford	Hanford	Tokai	Hanford
Computer codes used in evaluations[b]	MCNP/KENO	MCNP	MCNP	MCNP/KENO	MCNP/KENO	MCNP/KENO	MCNP/KENO	MCNP/KENO
Cross-section sets used in evaluations[c]	ENDF/B-V/IV	ENDF/B-V	ENDF/B-V	ENDF/B-V/IV	ENDF/B-V/IV	ENDF/B-V/IV	JENDL-3.2	ENDF/B-IV&V
Cross-section type[d]	cont/27grp	cont	cont	cont/27grp	cont/27grp	cont/27grp	cont/137grp	cont/27grp
Fuel compound[e]	oxide	oxide	oxide	oxide	oxide	oxide	oxide	oxide
Fuel compound form	solid	solid	solid	solid	solid	solid	solid	solid
Density of fuel[f]	86.7%	86.7%	86.7%	86.7%	86.7%	86.7%	55%	86%
Organization of fuel[g]	pins	pins	pins	pins	pins	pins	pins	pins
Cladding used for fuel[h]	Zirc-2	Zirc-2	Zirc-2	Zirc-2	Zirc-2	Zirc-2	Zirc-2	Zirc-2
Pu/(Pu+U) atom percent	1.51%	1.80%	1.80%	1.80%	2.01%	2.01%	3.03%	3.52%
U-235 atom percent	0.16%	0.71%	0.71%	0.71%	0.72%	0.72%	0.71%	0.71%
U-238 atom percent	99.84%	99.29%	99.29%	99.29%	99.28%	99.28%	99.29%	99.29%
Pu-238 atom percent	-	0.01%	0.01%	0.01%	-	-	0.50%	0.28%
Pu-239 atom percent	91.41%	91.84%	91.84%	91.84%	81.11%	71.76%	68.18%	75.39%
Pu-240 atom percent	7.83%	7.76%	7.76%	7.76%	16.54%	23.50%	22.02%	18.10%
Pu-241 atom percent	0.73%	0.37%	0.37%	0.37%	2.15%	4.08%	7.26%	5.08%
Pu-242 atom percent	0.03%	0.03%	0.03%	0.03%	0.20%	0.66%	2.04%	1.15%
Plutonium type as given in Table 1	WG	WG	WG	WG	FG	PG	PG	FG-PG
Shape of lattice[i]	cylinder	rectangle	rectangle	cylinder	cylinder	cylinder	rectangle	cylinder
Pitch of lattice	triangle	square	square	triangle	triangle	triangle	square	triangle
Number of experiments in each set	6	3	3	6	5	6	4	7
Fissile moderator used[j]	H$_2$O	H$_2$O	B-H$_2$O	H$_2$O	H$_2$O	H$_2$O	H$_2$O	H$_2$O
Reflector used	H$_2$O	H$_2$O	B-H$_2$O	H$_2$O	H$_2$O	H$_2$O	H$_2$O	H$_2$O
Maximum effective bias of experiments in set (Eff-Bias)	-0.0112	-0.0052	-0.0026	-0.0089	-0.0040	-0.0068	-0.0097	-0.0037
Maximum uncertainty of experiments in set (Exp-Uncer)	0.0080	0.0059	0.0045	0.0054	0.0061	0.0065	0.0051	0.0042
Exp-Uncer minus Eff-Bias (WCD)	0.0192	0.0111	0.0071	0.0143	0.0101	0.0133	0.0148	0.0079

a Definition of acronyms is MCT = MIX-COMP-THERM
b Codes MCNP[E-7] and KENO[E-8]
c ENDF/B-V/IV means cross-section set ENDF/B-V for MCNP and cross-section set ENDF/B-IV for KENO   JENDL-3 2 is cross-section set for both MCNP and KENO
d Cross-section type is either continuous cross sections (cont) or group cross sections (27grp, 137grp)
e Heavy metal is as an oxide
f MOX density given as percent of theoretical density taken as 11 00 g/cm^3
g Pins means organization of MOX is as pellets in fuel pins
h Zirc-2 means zircaloy-2 cladding
i Cylinder means shape of lattice is a cylinder   Rectangle means shape of lattice is a rectangle
j B-H$_2$O means borated water as moderator or reflector

Table E-2. Important Characteristics of Lattice Experiments with Weight Percent of Pu/(Pu+U) from 5% to 15% (from IHECSBE)

	MCT-003	MCT-003	MCT-012	MCT-012	MCT-012	MCT-012	MCT-012
Designation for experiments[a]	MCT-003	MCT-003	MCT-012	MCT-012	MCT-012	MCT-012	MCT-012
Facility where experiments conducted	WREC	WREC	Hanford	Hanford	Hanford	Hanford	Hanford
Computer codes used in evaluations[b]	MCNP	MCNP	MCNP/KENO	MCNP/KENO	MCNP/KENO	MCNP/KENO	MCNP/KENO
Cross-section sets used in evaluations[c]	ENDF/B-V	ENDF/B-V	ENDF/B-V	ENDF/B-V	ENDF/B-V	ENDF/B-V	ENDF/B-V
Cross-section type[d]	cont	cont	cont/238grp	cont/238grp	cont/238grp	cont/238grp	cont/238grp
Fuel compound[e]	oxide	oxide	oxide-poly	oxide-poly	oxide-poly	oxide-poly	oxide-poly
Fuel compound form	solid	solid	solid	solid	solid	solid	solid
Density of fuel[f]	94%	94%	N/A	N/A	N/A	N/A	N/A
Organization of fuel[g]	pins	pins	cubes, slabs	cubes, slabs	cubes, slabs	cubes, slabs	cubes, slabs
Cladding used for fuel[h]	Zirc-4	Zirc-4	plastic 471	plastic 471	plastic 471	plastic 471	plastic 471
Pu/(Pu+U) atom percent	6.63%	6.63%	7.60%	7.89%	14.62%	14.62%	14.62%
U-235 atom percent	0.71%	0.71%	0.15%	0.15%	0.15%	0.15%	0.15%
U-238 atom percent	99.29%	99.29%	99.85%	99.85%	99.85%	99.85%	99.85%
Pu-238 atom percent	-	-	0.59%	-	-	-	-
Pu-239 atom percent	90.65%	90.65%	67.97%	91.25%	91.42%	91.42%	91.42%
Pu-240 atom percent	8.55%	8.55%	22.95%	8.12%	7.97%	7.97%	7.97%
Pu-241 atom percent	0.76%	0.76%	5.57%	0.58%	0.57%	0.57%	0.57%
Pu-242 atom percent	0.04%	0.04%	2.92%	0.05%	0.04%	0.04%	0.04%
Plutonium type as given in Table 1	WG-FG	WG-FG	PG	WG	WG	WG	WG
Shape of lattice[i]	rectangle	rectangle	3D cube	3D cube	3D cube	3D cube	3D cube
Pitch of lattice	square	square	square	square	square	square	square
Number of experiments in each set	5	1	6	7	7	6	3
Fissile moderator used[j]	H_2O	$B-H_2O$	polystyrene	polystyrene	polystyrene	polystyrene	polystyrene
Reflector used	H_2O	$B-H_2O$	plexiglas	plexiglas	plexiglas	plexiglas	none
Maximum effective bias of experiments in set (Eff-Bias)	-0.0063	-0.0030	-0.0270	-0.0016	-0.0016	-0.0016	-0.0020
Maximum uncertainty of experiments in set (Exp-Uncer)	0.0071	0.0052	0.0058	0.0036	0.0027	0.0027	0.0037
Exp-Uncer minus Eff-Bias (WCD)	0.0134	0.0082	0.0328	0.0052	0.0043	0.0043	0.0057

a Definition of acronyms is MCT = MIX-COMP-THERM
b Codes MCNP[E-7] and KENO[E-8]
c ENDF/B-V is cross-section set for MCNP and KENO
d Cross-section type is either continuous cross sections (cont) or group cross sections (238grp)
e Heavy metal is as an oxide   Oxide-poly means mixture of MOX particles and polystyrene pressed into cubes and slabs
f MOX density given as percent of theoretical density taken as 11 00 g/cm³
g Pins means organization of MOX is as pellets in fuel pins   Cubes, slabs means organization of MOX-polystyrene is as cubes and slabs
h Zirc-4 means zircaloy-4 cladding   Plastic 471 means cladding is six mil plastic tape MM&M (3M) #471
i Rectangle means shape of lattice is a rectangle   3D cube means cubes and slabs stacked into the shape of a 3D-rectangular cube
j B-H₂O means borated water as moderator or reflector

Table E-3. Important Characteristics of Lattice Experiments with Weight Percent of Pu/(Pu+U) Greater than 15% (from IHECSBE)

	MCT-001	MCT-011	MCT-012	MCT-012
Designation for experiments[a]	MCT-001	MCT-011	MCT-012	MCT-012
Facility where experiments conducted	Hanford	Valduc	Hanford	Hanford
Computer codes used in evaluations[b]	MONK	MORET	MCNP/KENO	MCNP/KENO
Cross-section sets used in evaluations[c]	UKNDL	JEF2.2	ENDF/B-V	ENDF/B-V
Cross-section type[d]	cont	172gp	cont/238grp	cont/238grp
Fuel compound[e]	oxide	oxide	oxide-poly	oxide-poly
Fuel compound form	solid	solid	solid	solid
Density of fuel[f]	89.4%	94.2%	N/A	N/A
Organization of fuel[g]	pins	pins	cubes, slabs	cubes, slabs
Cladding used for fuel[h]	316 SS	Z3CND18.12 SS	plastic 471	plastic 471
Pu/(Pu+U) atom percent	19.70%	25.80%	30.00%	30.00%
U-235 atom percent	0.71%	60.15%	0.15%	0.15%
U-238 atom percent	99.29%	39.85%	99.85%	99.85%
Pu-238 atom percent	0.15%	-	-	-
Pu-239 atom percent	85.54%	89.00%	91.22%	91.22%
Pu-240 atom percent	11.46%	9.72%	8.13%	8.13%
Pu-241 atom percent	2.50%	1.21%	0.61%	0.61%
Pu-242 atom percent	0.35%	0.07%	0.04%	0.04%
Plutonium type as given in Table 1	FG	WG-FG	WG	WG
Shape of lattice[i]	rectangle	cylinder	3D cube	3D cube
Pitch of lattice	square	triangle	square	square
Number of experiments in each set	4	6	8	3
Fissile moderator used	H$_2$O	H$_2$O	polystyrene	polystyrene
Reflector used	H$_2$O	H$_2$O	plexiglas	none
Maximum effective bias of experiments in set (Eff-Bias)	-0.0103	-0.0006	-0.0018	-0.0086
Maximum uncertainty of experiments in set (Exp-Uncer)	0.0025	0.0017	0.0049	0.0052
Exp-Uncer minus Eff-Bias (WCD)	0.0128	0.0023	0.0067	0.0138

a. Definition of acronyms is MCT = MIX-COMP-THERM.
b. Codes MCNP,[E-7] KENO,[E-8] MONK,[E-9] and MORET[E-10]
c. ENDF/B-V is cross-section set for MCNP and KENO. UKNDL is cross-section set for MONK. JEF2.2 is cross-section set for MORET.
d. Cross-section type is either continuous cross sections (cont.) or group cross sections (172grp, 238grp).
e. Heavy metal is as an oxide. Oxide-poly means mixture of MOX particles and polystyrene pressed into cubes and slabs.
f. MOX density given as percent of theoretical density taken as 11.00 g/cm^3.
g. Pins means organization of MOX is as pellets in fuel pins. Cubes, slabs means organization of MOX-polystyrene is as cubes and slabs.
h. SS means stainless steel cladding. Plastic 471 means cladding is six mil plastic tape MM&M (3M) #471.
i. Cylinder means shape of lattice is a cylinder. Rectangle means shape of lattice is a rectangle cube. 3D cube means cubes and slabs stacked into the shape of a 3D-rectangular cube.

Table E-4.   Important Characteristics of Tank Experiments with Weight Percent of Pu/(Pu+U) to 31% (from IHECSBE)

	MST-001	MST-001	MST-001	MST-001	MST-001	MST-002	MST-003
Designation for experiments[a]	MST-001	MST-001	MST-001	MST-001	MST-001	MST-002	MST-003
Facility where experiments conducted	Hanford	Hanford	Hanford	Hanford	Hanford	Hanford	AWRE
Computer codes used in evaluations[b]	MCNP/KENO	MCNP/KENO	MCNP/KENO	MCNP/KENO	MCNP/KENO	MCNP/KENO	MONK
Cross-section sets used in evaluations[c]	ENDF/B-V/IV	ENDF/B-V/IV	ENDF/B-V/IV	ENDF/B-V/IV	ENDF/B-V/IV	ENDF/B-V/IV	UKNDL
Cross-section type[d]	cont/27grp	cont/27grp	cont/27grp	cont/27grp	cont/27grp	cont/27grp	cont/27grp
Fuel compound[e]	nitrate	nitrate	nitrate	nitrate	nitrate	nitrate	nitrate
Fuel compound form	liquid	liquid	liquid	liquid	liquid	liquid	liquid
Density of fuel[f]	1.31-1.68	1.31-1.68	1.31-1.48	1.70	1.31-1.68	1.09	1.11-1.52
Pu/(Pu+U) atom percent	22%	22%	22%	22%	22%	23%	30.7%
U-235 atom percent	0.70%	0.70%	0.70%	0.70%	0.70%	0.70%	0.72%
U-238 atom percent	99.30%	99.30%	99.30%	99.30%	99.30%	99.30%	99.28%
Pu-238 atom percent	0.03%	0.03%	0.03%	0.03%	0.03%	0.03%	-
Pu-239 atom percent	91.12%	91.12%	91.12%	91.12%	91.12%	91.12%	93.95%
Pu-240 atom percent	8.34%	8.34%	8.34%	8.34%	8.34%	8.31%	5.63%
Pu-241 atom percent	0.42%	0.42%	0.42%	0.42%	0.42%	0.45%	0.42%
Pu-242 atom percent	0.09%	0.09%	0.09%	0.09%	0.09%	0.09%	-
Plutonium type as given in Table 1	WG	WG	WG	WG	WG	WG	WG
Tank fissile liquid is in[g]	N/A	cylinder	cylinder	cylinder	cylinder	cylinder	slab
Auxiliary tank additional fissile liquid is in[h]	annular	annular	annular	N/A	N/A	N/A	N/A
Number of experiments in each set	2	5	2	1	1	1	10
Fissile moderator used[i]	soln H_2O	soln H_2O	soln H_2O	soln H_2O	soln H_2O	soln H_2O	soln H_2O
Reflector used[j]	B_4C-concrete	B_4C-concrete	poly-Cd cover	none	poly-Cd cover	H_2O	H_2O & poly
Maximum effective bias of experiments in set (Eff-Bias)	-0.0101	-0.0164	-0.0028	-0.0068	-0.0028	-0.0020	-0.0038
Maximum uncertainty of experiments in set (Exp-Uncer)	0.0016	0.0016	0.0016	0.0016	0.0016	0.0024	0.0025
Exp-Uncer minus Eff-Bias (WCD)	0.0117	0.0180	0.0044	0.0084	0.0044	0.0044	0.0063

a. Definition of acronyms is MST = MIX-SOL-THERM.
b. Codes MCNP,[E-7] KENO,[E-8] and MONK[E-9].
c. ENDF/B-V/IV means cross-section set ENDF/B-V for MCNP and ENDF/B-IV for KENO. UKNDL is cross-section set for MONK.
d. Cross-section type is either continuous cross sections (cont.) or group cross sections (27grp).
e. Heavy metal is as a nitrate dissolved in dilute nitric acid solution.
f. Solution density is in g/ml.
g. Containers for fissile solution are cylinders or slabs.
h. Annular tank surrounding central cylindrical tank or just an annular tank.
i. Soln H_2O means the moderator is the fissile nitrate solution.
j. B_4C-concrete means borated concrete. Poly-Cd cover means polyethylene reflector coated with Cd.

Table E-5. Important Characteristics of Tank Experiments with Weight Percent of Pu/(Pu+U) Greater than 31% (from IHECSBE)

Designation for experiments[a]	MST-004	MST-004	MST-004	MST-005	MST-005	MST-002	MST-001
Facility where experiments conducted	Hanford	Hanford	Hanford	Hanford	Hanford	Hanford	Hanford
Computer codes used in evaluations[b]	MCNP/KENO	MCNP/KENO	MCNP/KENO	MCNP/KENO	MCNP/KENO	MCNP/KENO	MCNP/KENO
Cross-section sets used in evaluations[c]	ENDF/B-V /IV	ENDF/B-V /IV	ENDF/B-V /IV	ENDF/B-V /IV	ENDF/B-V /IV	ENDF/B-V /IV	ENDF/B-V /IV
Cross-section type[d]	cont/27grp	cont/27grp	cont/27grp	cont/27grp	cont/27grp	cont/27grp	cont/27grp
Fuel compound[e]	nitrate	nitrate	nitrate	nitrate	nitrate	nitrate	nitrate
Fuel compound form	liquid	liquid	liquid	liquid	liquid	liquid	liquid
Density of fuel[f]	1.17-1.67	1.17-1.67	1.17-1.67	1.17-1.67	1.17-1.67	1.05	1.15-1.44
Pu/(Pu+U) atom percent	40%	40%	40%	40%	40%	52%	97%
U-235 atom percent	0.56%	0.56%	0.56%	0.56%	0.56%	0.70%	2.29%
U-238 atom percent	99.44%	99.44%	99.44%	99.44%	99.44%	99.30%	97.71%
Pu-238 atom percent	0.03%	0.03%	0.03%	0.03%	0.03%	0.03%	0.03%
Pu-239 atom percent	91.12%	91.12%	91.12%	91.12%	91.12%	91.12%	91.57%
Pu-240 atom percent	8.34%	8.34%	8.34%	8.34%	8.34%	8.34%	7.94%
Pu-241 atom percent	0.42%	0.42%	0.42%	0.42%	0.42%	0.42%	0.39%
Pu-242 atom percent	0.09%	0.09%	0.09%	0.09%	0.09%	0.09%	0.07%
Plutonium type as given in Table 1	WG	WG	WG	WG	WG	WG	WG
Tank fissile liquid is in[g]	cylinder	cylinder	cylinder	slab	slab	cylinder	cylinder
Auxiliary tank additional fissile liquid is in[h]	N/A	N/A	N/A	N/A	N/A	N/A	annular
Number of experiments in each set	3	3	3	3	4	2	3
Fissile moderator used[i]	soln H$_2$O	soln H$_2$O	soln H$_2$O	soln H$_2$O	soln H$_2$O	soln H$_2$O	soln H$_2$O
Reflector used[j]	none	H$_2$O	concrete	none	H$_2$O	H$_2$O	B$_4$C-concrete
Maximum effective bias of experiments in set (Eff-Bias)	-0.0060	-0.0048	-0.0024	-0.0114	-0.0026	-0.0020	-0.0032
Maximum uncertainty of experiments in set (Exp-Uncer)	0.0033	0.0033	0.0078	0.0036	0.0037	0.0024	0.0016
Exp-Uncer minus Eff-Bias (WCD)	0.0093	0.0081	0.0102	0.0150	0.0063	0.0044	0.0048

a. Definition of acronyms is MST = MIX-SOL-THERM.
b. Codes MCNP[E-7] and KENO[E-8].
c. ENDF/B-V/IV means ENDF/B-V for MCNP and ENDF/B-IV for KENO.
d. Cross-section type is either continuous cross sections (cont.) or group cross sections (27grp).
e. Heavy metal is as a nitrate dissolved in dilute nitric acid solution.
f. Solution density is in g/ml.
g. Containers for fissile solution are cylinders or slabs.
h. Annular tank surrounding central cylindrical tank.
i. Soln H$_2$O means the moderator is the fissile nitrate solution.
j. B$_4$C-concrete means borated concrete.

E-8

References

E-1. H. R. Dyer and C. V. Parks, "Recommendations for Preparing the Criticality Safety Evaluation of Transportation Packages," NUREG/CR-5661, April 1997.

E-2. J. J. Lichtenwalter, S. M. Bowman, M. D. DeHart, C. M. Hopper, "Criticality Benchmark Guide for Light-Water-Reactor Fuel in Transportation and Storage Packages", NUREG-6361, March 1997.

E-3. Nuclear Energy Agency, "International Handbook of Evaluated Criticality Safety Benchmark Experiments," Organization for Economic Co-operation and Development, NEA/NSC/DOC(95)03, September 2001 Edition.

E-4. B. L. Broadhead, C. M. Hopper, R. L. Childs, and C. V. Parks, "Sensitivity and Uncertainty Analyses Applied to Criticality Safety Validation," NUREG/CR-6655, Vols. 1 and 2 (ORNL/TM-13692/V1 andV2), U.S. Nuclear Regulatory Commission, Oak Ridge National Laboratory, November 1999.

E-5. B. T. Rearden and R. L. Childs, "Prototypical Sensitivity and Uncertainty Analysis Codes for Criticality Safety with the SCALE Code System," *Trans. Am. Nucl. Soc.*, Washington, D.C., November 2000.

E-6. M. E. Dunn and B. T. Rearden, "Application of Sensitivity and Uncertainty Analysis Methods to a Validation Study for Weapons-Grade Mixed-Oxide Fuel," 2001 ANS Embedded Topical Meeting on Practical Implementation of Nuclear Criticality Safety, Reno, NV, November 11–15, 2001.

E-7. MCNP is a three-dimensional, Monte Carlo radiation transport code that uses point-wise cross sections, developed by LANL. *MCNP – A General Monte Carlo N-Particle Transport Code*, Version 4B, Judith F. Briesmeister, Editor, Los Alamos Report, LA-12625-M, March 1997. Various versions of MCNP are available. This reference is for version 4B.

E-8. KENO is a three-dimensional Monte Carlo criticality module in the SCALE system that uses multigroup cross sections, developed by ORNL. *SCALE: A Modular Code System for Performing Standardized Analyses for Licensing Evaluations*, NUREG/CR-0200, Rev. 4 (ORNL/NUREG/CSD-2/R4), Vols. I, II, III, October 1995. Various versions of SCALE are available. This reference is for version 4.3.

E-9. MONK is a three-dimensional Monte Carlo radiation transport code that uses point-wise cross sections, developed by A.E.A Technology of the United Kingdom.

E-10. MORET is a three-dimensional Monte Carlo criticality code that uses multigroup cross sections, developed by C.E.A. of France.